YOUR KNOWLEDGE HAS VALUE

Bibliographic information published by the German National Library:

The German National Library lists this publication in the National Bibliography; detailed bibliographic data are available on the Internet at http://dnb.dnb.de .

Imprint:

Copyright © 2016 GRIN Verlag, Open Publishing GmbH
Print and binding: Books on Demand GmbH, Norderstedt Germany
ISBN: 9783668313446

This book at GRIN:

http://www.grin.com/en/e-book/341201/land-access-and-tenure-security-challenges-a-motivation-for-the-non-involvement

WUNI IBRAHIM YAHAYA, BARIKISA OWUSU ANSAH

Land access and Tenure security challenges. A motivation for the non-involvement of some youth in agriculture

GRIN Publishing

Land accessibility and tenure security challenges; the motivation for the non-involvement of some youth in agriculture: Investigating underlying issues in the Techiman Traditional Area

Department of Land Economy, Faculty of Built Environment, Kwame Nkrumah University of Science and Technology

April, 2016

Abstract: There is high unemployment among the youth in Ghana and across the world. With the educated youth, it is even higher despite the prospect employment opportunities in the agricultural sector. This leaves a lot of thinking in the minds of many as to what are the motivations for such attitude of the youth. This study hypothesized that land access and tenure security are the severe factors discouraging the youth from involving in agriculture in the Techiman Traditional Area.

The study critically undertook topical and policy interventions literature review and incorporated an empirical investigation of the hypothesis and underlying issues. The study employed both quantitative and qualitative data obtained largely from questionnaires and interviews surveys. A total of 200 youths who were not currently involved in agriculture were randomly selected from some communities in the Techiman Traditional Area. The study made use of the relative importance index scale to rank the factors that motivates the youth to stay away from agriculture and other data reporting techniques.

The survey indicated that land access and tenure security challenges are not the most severe factors discouraging the youth from participating in agriculture. The youth ranked it second as a factor that discourages them from agriculture with an RII value of **0.804.** The study found out that the youth perceived lack of interest and passion for agriculture to be the most violent factor that accounts for their non-involvement in agriculture and ranked it first with an RII value of **0.815**. They ranked the remaining factors in descending order from third to twelve as follows: Busy with other 'better jobs' **(0.652),** Lack of capital or funds **(0.579),** Schooling **(0.568),** Difficulty of the profession **(0.541)** , Negative perception about farming **(0.512),** Lack of relevant knowledge **(0.499),** Unreliable weather **(0.480)** and Unprofitability **(0.437).** The study also revealed that majority of the youth 140(60%) has future agricultural plans.

The study established that land access and tenure security challenges included high price of land, scarcity of agricultural land, customary restrictions, multiple sales of land, land racketeering, too short tenancy arrangement and refusal to release agricultural lands. And the study accordingly recommended the need for sustained integrated and holistic approaches to youth empowerment in agriculture.

Keywords: Youth, Agriculture, Land Access, Tenure Security, Empirical.

CONTENT

1.0 INTRODUCTION

Agriculture forms a core sector of the economy of Ghana. It employs a greater number of citizens of Ghana and contributes significantly to the Gross Domestic Product of Ghana. The Ghana Statistical Service (2013) identified that agriculture constitute about 36% of Gross Domestic Product (GDP) and provides myriad livelihood avenues for majority of Ghanaians especially the rural folks. It has been established by the Face of Agriculture (2011) that agriculture currently employs a greater number of youth in Ghana and anticipates that such would be sustained for a number of years. Ghana has a total land size of 24 million hectares and Ahwoi (2010) established that about 57% of the entire mass has been identified to be suitable for agriculture and as of 2009 an estimated 54% of these agricultural lands were under cultivation.

The youth forms a major proportion of the Ghanaian population and the future of Agriculture in the country depends on the involvement. The population of the youth has seen a remarkable increase for the past few decades and yet no such can be said about them who have joined the agricultural sector. The Ghana Statistical Service (2013) established that within the past five decades the population of the youth in Ghana has escalated from 1.1 million in 1960 to 2.3 million in 1984. And in 2000, the population increased to 2.3 million and that is obviously a sharp increase. The youthful population increased by 15% between 2000 and 2010. This sharp increase in the number of youth generates demand for employment and the non-existence of such will lead to mass unemployment in the case of Ghana. These youth include the skilled and unskilled, educated and uneducated, organized and unorganized among other (National Youth Policy, 2010). The Ghana Statistical Service (2013) revealed that the labour market receives about 300,000 youths every year and within which 3% is been absorbed by the formal sector and the remaining 97% of them survive in the informal sector or totally unemployed.

Survey evidence suggests that the active farming population is ageing in the country and the Techiman Traditional area of the Brong-Ahafo region is no exception. This must be addressed to facilitate sustainable food production and extension of food availability in the region from internal production. The Ghana Statistical Service (2013) established that mean age of farmers in Ghana is 45 years and the life expectancy in the Brong-Ahafo Region lies between 57-70 years. Accordingly, within the next 2 decades, young farmers do not take up or replace the ageing farmers or food producers, food production and security within the region will be seriously threatened.

In response to these future threats by the non-participation of young people in agriculture, the Government of Ghana has made several attempts to promote youth in agriculture as a way of tackling youth unemployment menace and the anticipated future food insecurity. Practically, these renewed commitment and effort by the Government to promote the youth in agriculture has not been successful and consequently their non-involvement in agriculture persist and this has hick-up unemployment concerns in recent times. The Food and Agriculture Organization (2011) conducted a study to unearth the reasons for the non-patronage of the youth to the programmes organized by the Government and other organizations to stimulate youth participation in agriculture and established that the non-involvement of the youth in agriculture were largely associated to the sector been made unattractive due to risk, cost, inefficacy, and its labor intensive nature.

However, there are strong indications to suggest that youth participation in the agriculture sector is very low. In view of the above, it is clear that even though agriculture has a potential of providing employment for the unemployed youth, the Ghanaian youth appear to be more attracted to formal sector employment and the non-existent 'clean jobs' and this attitude of the youth has resulted in great excess demand for jobs over supply of jobs.

Because access to land and tenure security are indispensable to the actual practice of agriculture, the study hypothesizes that land accessibility and tenure security challenges are the most violent motivations for the non-involvement of some of the youth in agriculture despite the menace of the unemployment in the country. There is the need for this research because while the idea sounds great in from the conceptual framework of research works perhaps in practice there might be other challenges that are more severe than land access and tenure security barriers.

The fact that these youth are not ready to undertake agriculture despite the great menace of unemployment in Ghana and with all the conscious and reasonable incentives available to encourage some of these youth to agriculture raises some important questions "why their non-involvement" and "how can we can promote agriculture as a feasible career for the youth in the study area?".

The paper then makes an empirical investigation into the reasons for the non-involvement of some of the youth in Agriculture in the Techiman Traditional area with a focus on land accessibility and tenure security challenges. The study also sought to explore other reasons that account for the non-involvement of some the youth in agriculture and to suggest practical ways of eradicating the problem. This forms the thesis statement or basis of the entire research paper.

2.0 REVIEW OF RELATED LITERATURE

2.1 Introduction

This chapter reviews literature on the non-involvement of some youth in agriculture. The review is composed of a description of youth, and proceeds with the description of agriculture, youth participation in agriculture, empirical studies on factors that accounts for the non-participation of some youth in agriculture and land accessibility. It also includes the state of Agriculture in Ghana, youth livelihood sources and the constraints of some of the youth in the Agricultural sector.

2.2 Youth

The term 'youth' connotes differently from country to country. Youth in the basic sense refers to the period between childhood and adulthood. It is said to be a period during which a person prepares himself to be a responsible and active member of society. In nearly all countries youth forms the active working population.

The National Youth Policy (2010) of Ghana opined that Youth is a period where a person transitions from being family dependent childhood into independent adulthood and actively integrates in the society as a responsible citizen. The United Nation rather defined the youth as persons aged between 15 and 24 years. The African Youth Charter (2006) of the African Union holds a different opinion. The African Youth Charter defines the youth as persons aged between 15 and 35 years.

The Youth Charter asserts that the term includes young people between the ages of 25-35. The definition of youth adopted for this study is the conglomeration of the definitions by both the United Nations and the African Youth Charter. For the purpose of this research youth shall be persons between the ages of 15 to 35 years. Mangal (2009) established that this age group is regarded as the most productive of any society as it contains people in the prime of their lives physically and mentally.

2.3 Agriculture

Basically, agriculture is defined as the production of goods and services and the rearing of farm animals to provide food, drugs, and chemicals among others to satisfy human wants. Agriculture in Ghana consists of a variety of agricultural products and is an established economic sector, and provides employment on a formal and informal basis. Ghana produces a variety of crops in various climatic zones which range from dry savanna to the wet equatorial forest and which run in east west bands across Ghana. Beyuo and Bagson (2013) established that agriculture is core to every nations development especially in this 21st century and consequently has been an investment focal point for both developed and the developing countries.

2.4 Involvement of Youth in Agriculture

Agriculture can be considered one of the livelihood pillars of any society as it is the sector that provides food to fulfill the basic need of feeding the people. In developing economies without or with little output from manufacturing and the service sector, agriculture and rural economic development are inextricably linked.

The United Nations and Food and Agriculture Organization (2014) rather opined that a very few young see a future for themselves in agriculture. This is a subject of investigation in this study. Youth are the major catalyst for change and a backbone of every nation and Valerie (2009) established that mobilizing them for national development through participation in agriculture is paramount. There are no definitive statistics from either the National Labour Commission or the Statistical Service on the involvement of youth in agriculture. However, there is compelling evidence of an ageing farmer population in the Ghana and this consequently gives indications that there is insufficient youth participation in agriculture.

In Ghana agriculture is mainly practiced by the old with an average age of farmers being 55 years and life expectancy averages about 55-60 years (Ghana Statistical Service, 2013). The import of this indication is that if young farmers do not replace the ageing farmers, the production of food and food security within the country will be seriously compromised in the next 10-15 years (Valerie 2009).

2.5 Constraints to youth involvement in Agriculture

The non-involvement of some youth in agriculture has galvanized the attention of several researchers, the government and international organizations. Valerie (2009) asserted that mobilizing the youth for national development by way of involvement in agriculture was indispensable. The Food and Agriculture Organization (2010) identified that the youth do not welcome agriculture because they see it as an occupation for old, illiterate and rural poor.

Despite attempts made by the Government, NGOs and private institutions to encourage youth participation in agriculture, the efforts have not reflected in the age distribution of the farming population. This could be as a result of constraints faced by young farmers and thus, making it very difficult to engage agricultural activities.

Agriculture being one of the foundation pillars of any society can only function as such if the insufficient youth participation is reversed (Mangal, 2009). Mangal stressed that there is insufficient youth participation in the agricultural sector even though this class of people is the most productive of any society as it contains people in the prime of their lives physically and mentally.

Akpan (2010) identified that the non-involvement of some of the rural youth could be traced to economic, social and environmental factors. His study established that inadequate credit facilities, low farming profit margins, and a lack of agricultural insurance, initial capital and production inputs were regarded the economic factors discouraging the youth. Social factors included negative public perception about farming and the attitude of parents encouraging youth to find other jobs to the neglect of agriculture. Environmental issues include land accessibility and tenure security concerns, continuous poor harvests, and soil degradation. The environmental factors identified by Akpan (2010) supports the hypotheses by the researchers that the non-involvement could be traced to land accessibility and tenure security challenges.

Adekunle *et al* (2009) also identified a number of constraints to the involvement of youth in agriculture. His study identified inadequate credit facility, poor returns to investment, non-existence of agricultural insurance, poor basic farming knowledge, insufficient access to tractors and traditional methods of farming, no ready market for smaller productions, farming being perceived to be difficulty even those who are not involved, negative perception about agriculture, insufficient initial capital, farmers are not respected, unprofitability of agriculture, continuous poor harvest, poor storage facilities, insufficient of land and soil degradation.

Adewale *et al* (2012) opined again that agriculture can be sustainably promoted by the youth because they have what it takes to promote agriculture but this is been constrained by their strong apathy towards it. Amadi (2012) supports the Adewale *et al* when his study established that the youth look down upon agriculture and thus classifying it a dirty occupation

According to White (2012), one reason why young people express reluctance to farm may reflect their aversion, not to farming as such, but to the long period of waiting that they face before they have a chance to engage in independent farming, even when land is available in the community. White stressed that in many or most agrarian societies the older generation –

parents, or community elders in places where land is controlled not individually but by customary law – retain land as long as possible. White is of the view that the tension between the desires of the older generation to retain control of family or community resources, and the desire of young people to receive their share of these resources, form their own independent farms and households, and attain the status of economic and social adulthood, is a common feature of agrarian societies.

White (2012) stipulated that to understand better the reasons behind why young people turn away from agriculture we need to take account of a number of problems, including: the deskilling of rural youth, and the downgrading of farming and rural life; the chronic government neglect of small-scale agriculture and rural infrastructure; and the problems that young rural people increasingly have, even if they want to become farmers, in getting access to land while still young.

Access to land is extremely important for young people trying to earn a livelihood in agriculture and rural areas. Land access is not only the number one requirement for starting farming, but it can also contribute to household food security and is a means for employment creation and income generation. The United Nations and Food and Agriculture Organization (2014) established that there is a limited access to land for the youth and the situation is doubled on the side of women youth. The only means for them to access land is through inheritance. The UN-HABITAT (2011) also supports the above when it was revealed from their study that in many parts of Africa, it is taboo for young people to access the family land while the parents are still alive.

The World Bank (2012) also shared the same view when it established that it can be especially difficult for young women to request enforcement of formal laws because they often lack the required knowledge, financial resources and confidence to protest against social norms and traditions. The United Nations and Food and Agriculture Organization (2014) again posited that even if youth do have access to land, they still need finance to cover the costs of planting and harvesting.

2.6 Interventions to Make Agriculture Attractive to the Youth

In response to the apathetic nature of youth participating to agriculture, a number of interventions became necessary to arrest the canker. Consequently, the Food and Agriculture Organization (2011) observed that it is imperative to encourage the number of producers in agriculture and the youth population is strategic to the success of these efforts.

Kirui *et al* (2010) identified that transforming agriculture from subsistence level to commercial farming can motivate youth to engage in agriculture. This will not only aid in achievement of Vision 2030 that positions agricultural sector as a key driver for delivering the 10% annual economic growth, but will provide employment for the unemployed youth. Kangai *et al* (2011) however stressed that such motivations can be possible through increasing productivity, commercialization and competitiveness of agricultural commodities and enterprises in order to make agriculture more attractive to the youth.

Cotula (2011) rather believes the government has the biggest role to play in this regard. Accordingly he opined that the government should ensure that arable government land is only used for agricultural purpose, fairly distributed among young male and female farmers and that mechanisms to be put in place to help youth have sustainable agriculture. These could be achieved through promotion of land reforms and creation of laws that ensure young people's access to production resources that ensure equal opportunities for young men and women should be adopted. Food and Agriculture Organization (2011) agrees with Cotula (2011) when their study observed that the government can adopt laws and public policies relevant to young rural people and small producers. Such laws and policies should facilitate access to credit by the youth and reduce inequalities in rural areas to ensure young people's access to land. They should also provide young men and women with future prospects in farming and strengthen their identity.

Mcnulty and Nagarajan (2005) also identified government intervention as way of promoting youth in agriculture. To them, the government ought to implement legally binding consultation mechanisms with rural communities and rural youth movements while drafting policies related to productive resources. Such policies should respect land resources and its natural production cycles, and guarantee a healthy and sustainable environment for future generations (Mcnulty & Nagarajan, 2005). Increasing the availability of land for agriculture by stopping the illegal miners from use of the land can be an intervention by the government to promote youth in agriculture Herbel *et al* (2010).

Food and Agriculture Organization (2011) established that youth customarily inherit small plots of land and lack access to finance to acquire additional land. It was also established in their study that in India, cooperative farming has proved to be successful in overcoming this constraint a phenomenon that Ghanaian youths can borrow in order to improve their level of participation in agriculture

Herbel *et al* (2010) opined that agribusiness centers with storage and processing facilities should be created for young farmers in order to link farmers and traders and to act as a venue for training, sensitization and capacity building programmes on market actors, financing opportunities and new agricultural technologies. The Food and Agriculture Organization (2009) rather opined that the governments, together with funding agencies, should facilitate the development of infrastructure especially roads linking young producers from rural areas to markets.

The International Fund for Agricultural Development (2009) established that marketing of agricultural products, agricultural technologies and financial management in agriculture should be included in educational curricula from the secondary school onwards. The government should also provide farmers and indigenous people with the necessary conditions for competitive and sustainable marketing of their agricultural products (IFAD, 2009).

3.0 RESEARCH METHODOLOGY AND MATERIALS

This component presented the research methodology of the study. It described and justified the methods and processes that were used in order to collect data in answering the research questions. The chapter focused on the research design adopted, sampling techniques, sample frame and size the key study variables, the various tools and techniques employed in gathering the data and units of analysis. This section further considered the methods of data collection as well the methods that were adopted in the data processing, analysis and reporting.

The investigation adopted a survey instrumentation to study the variables concerned. In the study area, the unit of analysis was the individual youth who were not involved in agriculture. The sampling techniques that were judiciously employed to collect the data included purposive sampling technique and snowball sampling technique. Under the Purposive sampling technique, the researchers based on their knowledge and understanding of the population handpicked certain individuals for their relevance and suitability to the subject of investigation.

In this study, the sample frame consisted of all the youth who were not involved in agriculture in the Techiman Traditional Area in the Brong-Ahafo Region. This was not actually used to collect the data because there was no available data or record containing the list of all the youths who were not involved in agriculture in the Techiman Traditional Area. That motivated the extensive use of the purposive sampling technique to meet the sample size during the data collection. A Sample of 200 youth who were not involved in agriculture in the Techiman Traditional area were then surveyed using the purposive sampling technique and snowball sampling techniques from the potential respondents who were traced for the study.

The sources of data for the study include primary sources and secondary data. The primary data included the data that were collected directly from the respondents. And the secondary data were from research articles, journals, research papers, thesis among others. The data collection tools and instruments included the use of questionnaires and interviews. The questionnaires included both close and open ended questions and such was used to collect the data directly from the respondents. The interview guide was also used to collect the primary data where necessary from some other youth who were not covered under the questionnaires. Pre-testing of the research instruments (questionnaires) for the target respondents were carried out to check internal consistencies, ambiguities and clarity.

The Questionnaire comprised only three sections: the first section solicited demographic data; the second part elicited views on the relative importance of the factors that accounts for the non-involvement of some of the youth in agriculture. These were the factors that were identified from the literature review as constraints to youth participation in agriculture. The Questionnaire was administered to the 200 youths. The third section investigated land accessibility concerns, future agriculture prospect for the youth and underlying issues. A total of five (5) youth who were not involved in agriculture were taken through the questionnaire during the pre-testing. Revisions were subsequently made to some of the questions based on the findings from the pre-testing.

The method of data analyses that were employed in this report is a descriptive statistical technique of data analysis involving the creating of summary values. It involved the critical examination and explanation of data collected during the fieldwork. Tables, charts, percentages, Relative Importance Index Analysis and graphs were used in the case of the quantitative technique, while descriptions were used in the case of the qualitative analysis. After collecting the data, all questionnaires were cross checked and edited for mistakes. The questionnaires were then numbered and coded carefully. This was done to facilitate data entry to the computer and rectification of errors. The responses from the questionnaires were then entered into a template prepared using the Statistical Package for Social Sciences (SPSS) for further processing.

The factors that accounted for the non-involvement of some of the youth in agriculture were analyzed using the Relative Importance Index (RII) scale to quantify and rank the factors.

Accordingly, the Relative Importance Index (RII) was employed. RII was used for the analysis because it best suited the purpose of this study. On each of the factors the youth were asked to indicate the quantitative extent to which that variable were either a factor motivating the non-involvement based on a five-point scale where: 1 **meant** not severe, 2 **meant** less severe, 3 **meant** Neither severe, 4 **meant** severe and 5 **meant** very severe. This was carried out in order to empirically quantify and rank the extent to which each factor contributes to the non-involvement of the youth in agriculture. In the calculation of the Relative Importance Index (RII), Badu *et al* (2013) established that the formula was $RII = \frac{\Sigma W}{A*N}$. Where, **W** represented weighting given to each statement by the youth and ranged from **1** to **5**; **A** represented the highest response Integer **(5)**; and **N** represents the total number of respondents

4.0 RESULTS AND DISCUSSION

4.1 Introduction

This component concisely explored the background of respondents to serve as confounding variables. This was relevant because, the demographic features of the respondents facilitated further understanding of the quantum of the variables under the study. The sample characteristics have links with the issue under study. For this paper, the demographic data of emphasis included gender, age, educational status of the respondents, marital status and community membership status of the respondents.

4.2 Demographic Features of the Respondents

4.2.1 Ages of the Respondents

The ages of the youth has a direct link on how they perceive agriculture. The researchers been aware of this included the ages of the respondents to investigate the impacts of the various age brackets. For this study, the youth were categorized into 15-20years, 21-25years, 26-30years and 31-35years. Such categorization was motivated by the age range of the youth defined in the literature review. Out of the 200 respondents 40 of them were between the ages of 15-20years, 48 of the respondents were between the ages of 21-25years, 90 of the respondents were between the ages of 26-30years.The remaining 22 of the respondents were between the ages of 31-35years. **Table 1** shows the ages of the respondents.

4.2.2 Gender of the Respondents

The researchers believed gender of the respondents have an inextricable link with the involvement in agriculture. This is so because there is gender disparity in the accessibility of land in Ghana. The survey data revealed that there were more of the males (130) than their female (70) counterparts who were not involved in agriculture. **Figure 1** below is a pie chart showing the gender of the respondents.

4.2.3 Educational Status of the Respondents

The researchers are optimistic that there is a link between educational attainment of the youth and the perception about agriculture. In order to study the relationship between educational status and the underlying issues, the researchers categorized the educational status of the respondents into none, basic, secondary, diploma as well as degree and more. The data revealed that 40 of the respondents had no basic education at all and the remaining 160 respondents had at least basic education. The import of this is that most of the youth in the Techiman traditional area have at least basic education. **Table 2** below shows the educational status of the respondents.

4.2.4 Marital Status of the respondents

The researchers also wanted to investigate into whether or not marital status of the respondents has a link with the non-involvement in agriculture and underlying issues. The researchers therefore accordingly grouped the youth into single, married and divorced. The survey data established that most of the youth were single (155) and a very few were married (45) and none was divorced at all. This was logical because most of the youth were within the age bracket 15 and 30 years (178 respondents) and marriages in the community are delayed because of education. **Figure 2** below is a bar chart showing the marital status of the respondents.

4.2.5 Community Membership Status of the respondents

Land accessibility for agriculture has a link with the community membership status of the respondents. The researchers been indigenes of the community are much aware of the influx of northerners to area in search of jobs including farming. The researchers therefore categorized the respondents into migrants and indigenes for the study. The survey established that there were more migrants (135) than indigenes (65) and that had a bearing on land accessibility and tenure security. **Figure 3** below is a pie chart showing the community membership status of the respondents

Table 1: Ages of the respondents

Age Bracket	No. of Respondents	% of Respondents
15-20	40	20
21-25	48	24
26-30	90	45
31-35	22	11
Total	**200**	**100**

Source: Field Survey, 2016

Table 2: Educational Status of the Respondents

Educational Status	No. of Respondents	% of Respondents
None	40	20
Basic	64	32
Secondary	76	38
Diploma	15	7.5
Degree and more	5	2.5
Total	**200**	**100**

Source: Field Survey, 2016.

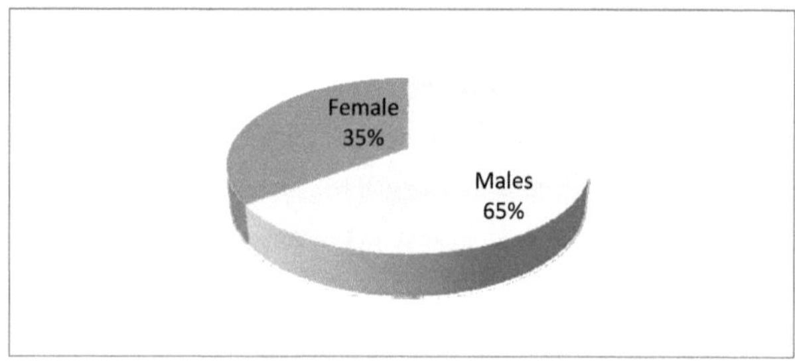

Figure 1: Gender of the Respondents

Source: Field Survey, 2016

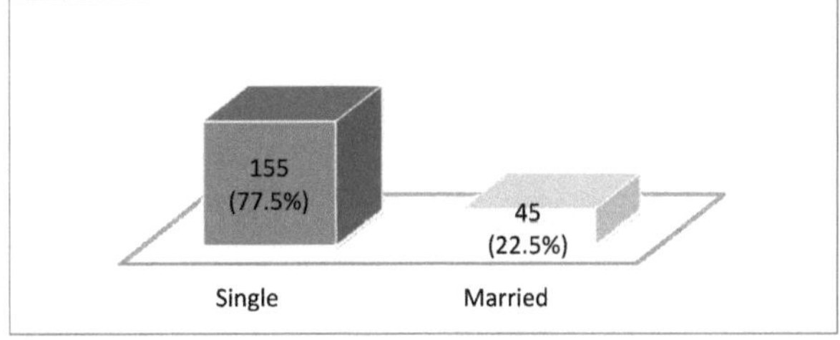

Figure 2: Marital status of the Respondents

Source: Field Survey, 2016.

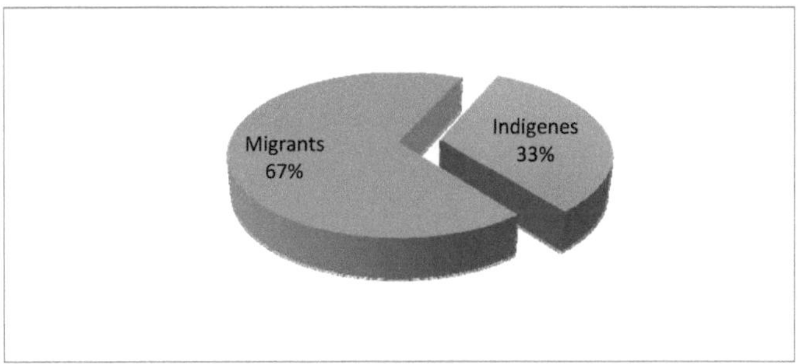

Figure 3: Community membership Status

Source: Field Survey, 2016.

4.3 Factors and Motivations for the Non-Involvement of Youth in Agriculture

The researchers hypothesized that land accessibility and tenure security challenges are the most significant motivation for the non-involvement of some of the youth in agriculture in the Techiman Traditional area. The researchers are also well-convinced that the findings of other researchers on the constraints of youth participation in agriculture are very significant. It was therefore appropriate to subject the some of the factors to relative importance index scale test to find out in the opinions of the youth which one is more significant.

Table 3: motivations for youth non-involvement in agriculture

Cause or Motivation	1	2	3	4	5	W	RII	Rank
Lack of interest and passion	20	15	20	20	125	815	0.815	1
Busy with other 'better jobs'	33	32	30	60	45	652	0.652	3
Lack of capital or funds	45	50	1	89	15	579	0.579	4
Difficulty of the profession	22	100	18	35	25	541	0.541	6
Unprofitability	75	31	78	14	2	437	0.437	10
Schooling	44	68	32	28	36	568	0.568	5
Lack of relevant knowledge	20	112	6	33	19	499	0.499	8
Land access and security challenges	19	24	14	20	123	804	0.804	2
Unreliable weather	78	42	3	76	1	480	0.480	9
Negative perception about farming	65	34	40	46	15	512	0.512	7

Source: Field Survey, 2016.

The table above shows the scales, weights, RII values and the corresponding rankings of the various factors that motivate the youth from undertaking agriculture. The higher the RII value or the closer an RII value to 1.000 means the youth perceives it to be a serious factor that accounts for their non-involvement and the smaller the RII value or the closer of an RII value to 0.000 means the youth do not perceived it to be a critical factor that the accounts for their non-involvement in agriculture. The study excluded the analysis of factors with RII less than 0.500 because, such factors are perceived to be of less importance as a factor discouraging the youth from undertaking agriculture. The relative importance index scale reveals the following.

4.3.1 Lack of interest and passion for agriculture

Counting on the exuberance of the youth, they undertake and pay attention to activities and jobs they are passionate about with or without motivations. To some of these youth nothing at all could be done to encourage them to take up a career in agriculture since that is not their field. Lack of interest and passion for agriculture was ranked first as a factor that motivates the youth to stay away from agriculture with an RII value of **0.815**; which is very close to

1.000. The import of this is that most of the youth are not involved in agriculture because they are really not interested in the field.

4.3.2 Land access and tenure security challenges

Land is an indispensable tool for agricultural activities since invariably all activities in the entire universe are carried on land. Most youth do not own land due to a number of restrictions and setbacks. Land in the area can be acquired either through inheritance, gift, lease, tenancy agreement or purchase. The youth ranked land access and tenure security challenges second as a factor discouraging them from undertaking agriculture with an RII value of **0.804;** which is also close to 1.000 indicating that the youth actually perceive it to be of great impact. The findings however, are contrary to the hypothesis of the researchers that land access and tenure security challenges were the main factor discouraging the youth. The study considered and furthered the analysis with regard to land access and tenure security challenges and the other factors that account for the non-involvement of some youth in agriculture.

4.3.3 Busy with other better jobs

The youth ranked 'busy with other better jobs' third as a factor that accounts for their non-involvement in agriculture with an RII value of **0.652.** The youth find agriculture as a job that pays less and involves a lot of work. They also believe that agriculture or farming is a dirty job and that condition of work is deplorable. They tend to seek and acquire jobs that have better conditions of work and with better pay. This is found to be the ideological orientation of the educated youth.

4.3.4 Lack of capital or funds

Agricultural activities in the area are associated with heavy sums of money. This is more severe where the youth would have to acquire the land through a lease or purchase. They would also have to purchase farm implements including cutlasses, fertilizers, among others. These collectively coupled with others require that a youth wanting to undertake agriculture should have some level of funding and this has been a great setback since most of the youth have no other means of getting money. And besides, the area is a low income community. Consequently, the youth ranked lack of capital as a factor that accounts for their non-involvement with an RII value of **0.579.**

4.3.5 Schooling

Some of the youth indicated that they were not in agriculture because they were still schooling. Others indicated that they had plans furthering their education. But one can argue that some crops are annual crops and others such as the food crops take less than a year to bear fruits. For reasons that are best known to the youth, they are not ready to undertake agriculture within such short periods. They therefore ranked schooling fifth with an RII value of **0.568**

4.3.6 Difficulty of the profession and Negative perception about farming

The youth ranked difficulty nature of agriculture and negative perception about farming were ranked sixth and seventh as factors that account for the non-involvement of some youth in agriculture with RII values of **0.541** and **0.512 respectively.** A larger segment of the Techiman Traditional Area still make use of the traditional method of farming using simple hoe and cutlass and thus requiring the direct human involvement of the youth. This is found to be tiresome and discouraging. The youth also perceives agriculture to be for the illiterate, old or rural folks as identified by Kpan (2010). This perception has been heightened into a great hatred for agriculture and consequently their non-involvement.

4.3.7 Lack of relevant knowledge, Unreliable weather and Unprofitability

Lack of relevant knowledge in agricultural activities, unreliable weather and unprofitability were ranked by the youth as eighth, nine and tenth factors accounting for the non-involvement of some youth in agriculture with RII values of **0.499, 0.480** and **0.437 respectively.** These were however not analyzed into their RII values were less than **0.500.**

4.4 Agricultural Experience of the Respondents

4.4.1 Past involvement in agriculture

Agriculture especially farming activities is the most predominant economic activity of most people in the Techiman Traditional Area. The analysis therefore considered the past farming experience of the respondents in order to ascertain whether or not some of the respondents have ever been involved in agricultural activities and if yes why did they stop farming. It was noticed that 120 (60%) of the respondents said that they have ever involved in farming activities for a livelihood and 80 (40%) of the respondents said they have never involved in any farming activities for a livelihood. The table below shows the responses of the respondents to the question *'have you ever involved in agriculture?'*

Table 4: Have you ever involved in agriculture?

Responses	No. of response	% of responses
Yes	120	60
No	80	40
Total	**200**	**100**

Source: survey data, 2016.

4.4.2 Land ownership and acquisition methods

To understand the extent to which land access plays a role in their non-involvement, questions on land ownership and accessibility were administered to the respondents who had past experience with agriculture. The table below shows the land ownership status of the respondents who had past farming experience.

Table 5: Do you own or have access to any land?

Response	No. of Responses	% of Responses
Yes	45	37.5
No	75	62.5
Total	**120**	**100**

Source: Survey Data, 2016.

The analysis revealed that a fewer youth 45 (37.5%) own land or have access to land in the community and majority of the youth 75 (62.5%) do not have their own land. This is perhaps due to the challenges associated with land access and tenure security.

4.4.3 Methods of land access

The study further investigated within the 45 respondents who owned land to identify the methods by which they acquired their lands. As a matter of established customary practices, land in the Techiman traditional area can be acquired predominantly through inheritance, gift, and tenancies, purchased, and hired. Agricultural land in the area is very scarce as most of the lands are being converted into other uses. The table below shows the methods of land access and acquisition by the respondents

Table 6: methods of land accessibility and acquisition

Mode of access	No. of Responses	% of Responses
Inheritance	20	44.44
Gift	5	11.11
Hiring	9	20.00
Purchased	2	4.44
Tenancy	9	20.00
Total	**45**	**100**

Source: Field Survey, 2016.

The survey indicated most of the youths acquire land through inheritance 20 (44.44%). The system of inheritance in the area is matrilineal. Some of the youth also acquired the lands through hiring and Tenancy 9 (20% each) from other people and family members. The least of them acquired their lands through purchase 2 (4.44%) because land prices are very high in the area.

4.4.4 Reasons why they stopped farming

The nucleus of the study is to investigate the reasons for the non-involvement of some of the youth in agriculture with a focus on land access and tenure security challenges. Having identified that a sizable of the youth had past agricultural experience, the study sought to explore the factors that motivated them to stop agricultural. The respondents who had agricultural experience were surveyed to solicit the reasons why they stopped farming after they were gainfully employed in agriculture. The reasons why they stopped farming were then identified **(Table 8)**.

Table 7: Reasons why some of the youth stopped farming

Motivations	No. of Responses	% of Responses
Land access challenges	24	20.00
Difficult and laborious	32	26.67
Unprofitability	8	6..67
Acquisition of other jobs	10	8.33
Unreliable weather conditions	5	4.17
Financial problems	15	12.50
Lack of knowledge	5	4.17
Bad market for produce	8	6.67
Desire to further education	10	8.33
Crops failure	3	2.50
Total	**120**	**100**

Source: Field Survey, 2016.

The data from the table revealed majority of the youth were discouraged from farming because they understood agriculture to be difficult and laborious 32 (26.67%). The impact of this is aggravated because such youth have no interest and passion for the activity and hence had no motivation to stay. A significant number of the respondents were discouraged by land access constraints 24 (20.00%) and financial problems 15 (12.50%). Some these youth will want to get enough money within the shortest possible time and so decided to expand their farms but such intentions and aspirations were thwarted by land accessibility challenges and lack of funds. Some of the youth had other jobs 10(8.33%) that they perceived were better agriculture and others had to further their education 10(8.33%) because they had joined agriculture to raise funds to cater for their education. Other were also discouraged by unprofitability, unreliable weather conditions, lack of knowledge on good agricultural practices, bad market for produce and crops failure as indicated in the table above.

4.5 Land Accessibility and Tenure Security Challenges

As part of the focus of the study, the study sought to identify the practical challenges associated with land accessibility and tenure security challenges. The questionnaire solicited views on some the challenges that the youth face in their attempt to access land and the tenure security. All the respondents were asked to indicate the practical setbacks associated with the land accessibility and tenure challenges. Below are the responses of the respondents (**Table 7**)

Table 8: land accessibility and tenure security challenges

Challenges	No. of Responses	% of Responses
High price of land	47	23.50
Scarcity of agricultural land	33	16.50
Customary restrictions	12	6.00
Multiple sale of land	54	27.00
Land Racketeering	25	12.50
Too short tenancy arrangements	20	10.00
Refusal to release agricultural lands	9	4.50
Total	**200**	**100**

Source: Field Survey, 2016.

The responses from the table above indicate that there are indeed a numbers of challenges in land access. Majority of the respondents indicated that there are rampant cases of multiple sale of land 54(27.00%) and that sometimes, in an attempt to purchase land, one might end up purchasing litigation or encumbrances. A couple of the respondents 47(23.50%) indicated that the price of land either by purchase or lease is very high. Others identified that some of the challenges include scarcity of agricultural land 33(16.5%), Land Racketeering 25(12.5%) and too short tenancy arrangement 20(10.00%). A number of less significant challenges included customary restrictions 12 (6.00%) for the female counterparts and Refusal to release agricultural land 9(4.50%)

4.6 Future Agricultural Plans

In order to appreciate the lifelong relationship between the youth and agriculture, the study sought to identify whether or not the concerned youth had plans going into agriculture as future careers. The responses indicated that greater proportions 140 (70%) of the youth had future agricultural plans, 60 (30%) indicated they had no future agricultural plans. This is contrary to the position by the previous researchers that future food security in the region is been threatened. The results from this survey have policy implication and the researchers maintain a pessimistic view that future food security can be improved if some of the discouraging factors are eliminated. The responses of the respondents as to their future agricultural plans are indicated below **(Table 9)**

Table 9: Do have plans going into Agriculture as a future career?

Responses	No. of Responses	% of Responses
Yes	140	70
No	60	30
Total	**200**	**100**

Source: Field Survey, 2016.

4.7 Reasons for not having future agriculture plans

The researchers further investigated the reasons why some of the youth had no plans of going into agriculture as a future career. This was included to augment the investigation of the overall reasons why some of the youth are not involved in agriculture. A number of reasons were advanced as to why they had no future plans of going into agriculture as a career **(Table 10)**. Most of them still maintained that agricultural activities using the traditional methods are difficult and laborious 15 (25%). A significant number indicated that they had no plans because they have no interest and passion 14(23.33%) for agriculture. Others indicated that they wanted to be in the formal sector 10(16.67%) and some maintained they had better jobs 10(16.67%) elsewhere. Only 6(10%) of 60 youth were discouraged by land problems as can be seen in the table below

Table 10: Reasons for not having future agricultural plans

Reason	No. of Responses	% of Responses
Difficult Nature	15	25.00
Not Interested	14	23.33
Land problem	6	10.00
Desire to pursue Academia	10	16.67
Unreliable weather	5	8.33
Acquisition of other jobs	10	16.67
Total	60	100

Source: Field Survey, 2016.

5.0 SUMMARY AND CONCLUSIONS

They study hypothesized that land accessibility and tenure security challenges were the major factors accounting for the non-involvement of some youth in agriculture in the Techiman Traditional Area. The analysis using the Relative Importance Index Scale did not confirm the hypothesis entirely when the youth ranked it second, a factor discouraging them from undertaking agriculture. It was established in the study that lack of interest and passion is the most severe factor discouraging the youth from agriculture. Other significant factors included busy with other better jobs, lack of capital or funds, schooling, difficulty of the profession and negative perception about farming, lack of relevant knowledge, unreliable weather and unprofitability.

It was however identified that a number of the youth have agricultural experience 120(60%) and stopped due to a number of reasons closely related to the reasons why they are not involved in agriculture. The mode of accessing land was predominantly through inheritance and the most affected in this direction are the migrant youth. The land access and tenure security in the area was also bedeviled with high price of land, scarcity of agricultural land. Customary restrictions, multiple sale of land, land racketeering, too short tenancy arrangements and refusal to release agricultural lands.

The study identified that majority of the youth 140 (70%) have future plans of going into agriculture. The import of this is that it is contrary to the conclusions from the previous studies that future food security in the area is threatened. The researchers however maintain a pessimistic view that future food security is not threatened. A fewer youth had no future agricultural plans 60(30%) and were motivated by difficult nature of agriculture, lack of interest and passion for agriculture, land access problems; desire to pursue academia, unreliable weather and acquisition of other jobs.

RECOMMENDATION

From the findings of the study, the researchers make the following policy interventions. In the first place, district assembly and all stakeholders should make sustained effort to galvanize the interest and passion of the youth towards agriculture. These can be achieved by acknowledging successful youth farmers on farmer's day celebration. These youth should be awarded and constantly be motivated to stay in the occupation. The youth will get motivated and be passionate about agriculture if they have great motivation beyond getting income. Those youth who happen to find themselves in agriculture should also be motivated to stay in the field.

In addition to the above, access to land and tenure security should be improved. These can be achieved by encouragement deed registration on the tenure period of acquired lands through tenancy and share cropping. The ministry of youth and agriculture in collaboration with the chiefs, district assemblies, and all stakeholders should device sustainable land access and tenure security policy interventions that are gender insensitive to make available land available and secure to them in undertaking agriculture.

Moreover, improving access to funds can be of great impact. Most of the youth are unemployed and do not also have land on their own. This is aggravated by their low level of income. Most financial institutions view the unemployed youth as risky clients for borrowing monies because they do not have assets to be used as collateral security and this has hardened the financial situation of the youth. Financial Institutions should develop agriculture financing packages and loan products that target and favor the youth who are seen as high risk clients. When funds are made easily accessible by credit facilities and agricultural banks, the youth can acquire lands, machinery and the logistics needed to stay in agriculture.

Furthermore, the method of practicing agriculture in the area is still the traditional approach involving the use of the hoes, axes, Wellington boot and cutlass. This is labour intensive and the youth find the occupation so difficult. This anomaly can be eliminated if the government through the Ministry of Agriculture and Ministry of Youth make sustained effort to mechanize agriculture and the youth should also be supported with funds to purchase tractors, fertilizers, bulldozers, combined harvester and their prices should be subsidized from the youth.

Policy interventions to curb land access and tenure security challenges should include enforceable punitive regulations and mechanisms to offenders who are guilty of multiple sale of land and land racketeering. Tenancy agreement from the land owners should be made relatively long to enable extended land access for farming and land owners should be encouraged by chiefs, district assembly, assemblymen and all stakeholders to release agricultural lands for farming.

Finally, agriculture (farming) is used in the basic and junior high schools as punishment mechanism for law breakers and the even the course module for agriculture is not even a separate discipline of study at the basic level. a negative mindset is therefore painted in that regard for agriculture and worsened by the lack of relevant knowledge in the practice of agriculture. Therefore, agricultural sensitization should be embarked sustainably at the community level and basic schools to create a good image for the field. The youth should be educated on good and sustainable agricultural practices through open forums, talks, symposiums, and seminars from the qualified resources persons and experts in the field of agriculture.

In a nutshell, the study recommends that policy interventions should adopt integrated and holistic approaches to empower youth in agriculture instead of employing piecemeal approaches. The study identified that some of the youth have plans of going into agriculture in the future and this indicates that they still have a little passion for agriculture. The Ministries of Youth and Agriculture in collaboration with Youth Advocacy Organizations should make sustained effort in that regard to empower the youth in agriculture.

6.0 REFERENCES

o Adekunle, O. A. *et al* (2009). *Constraints to Youth's involvement in gricultural production in kwara state*, Nigeria. Journal of agricultural extension, vol. 13(1), Pp 102-108.

o African Youth Charter. (2006). *The role of youth in its popularisation and ratification. Prepared by: Dabesaki Mac-Ikemenjima.*

o Ahwoi,, K. (2010). *Government"s Role in Attracting Viable Agricultural Investment: Experiences from Ghana, Paper presented at the World Bank Land Governance Conference, 26-27 April.* Washington DC.

o Akpan, B.S. (2010). *Encouraging Youth's Involvement in Agricultural Production and Processing.* International Food and Policy Research Institute, Nigeria

o Amadi, P.N.U. (2012). *Agricultural Entrepreneurship Development for Youth Empowerment in Nigeria: Constraints and Initiatives for Improvement(Vol.2 (10).* Journal of Educational and Social Research.

o Badu *et al.* (2013*) Rural Infrastructure Development in the Volta Region of Ghana: Barriers and Interventions.* Journal of Financial Management of Property and Construction,18, 142-159. http://dx.doi.org/10.1108/JFMPC-11-2012-0040

o Beyuo. A. N. and Bagson. E. (2013). *Youth in agriculture: prospects and challenges in the sissala area of Ghana.* UDS. Wa

o Cotula, L. (2011). *The Outlook on Farmland Acquisitions. International Institute for Environment, and Development (IIED),* Policy Brief, March 2011.

o FAC, (2011) *Future Framers: Exploring Youth Aspirations of African Agriculture Future Agriculture Consortium Policy brief 037.* www.future-agricultures.org

o FAO. (2011). *Agriculture in Ghana: Facts and Figures (2010):* Statistics, Research and Information Directorate (SRID). Accra.

o FAO. of the United Nation. (2014). *Youth and agriculture: Key challenges and concrete solutions.* HEADQUARTERS: FAO of United Nations.

o Ghana Statistical Service (2013). *2010 Population & Housing Census National Analytical Report.* Accra Ghana.

o Herbel, D. et al (2010). *Good Practices in Building Innovative Rural Institutions to Increase Food Security,* FAO, 2010.

o IFAD, FAO and WORLDBANK. (2009). Gender in agriculture Sourcebook.

o Kangai, E. et al (2011). *Incentives and Constraints of Financing Mechanisms for Compliance to Global GAP Standards among Smallholder Horticultural Farmers in Kenya. Presented in AGRO 2011 Biennial Conference, College of Agriculture and Veterinary Sciences,* University of Nairobi, Kenya, 26th - 28th September, 2011.

o   Kirui, O. et al. (2010). *Awareness and Use of M-banking Services in Agriculture: The Case of Smallholder Farmers in Kenya: Paper presented at the Joint 3rd African Association of Agricultural Economists (AAAE) and 48th Agricultural Economists Association of South Africa (AEASA) Conference,* Cape Town, South Africa, September 19-23, 2010.

o   Mangal, H. (2009*). Best Practices for Youth in Agriculture*: The Barbados, Grenada and Saint Lucia Experience. Final report.

o   Mcnulty, M. and Nagarajan, G. (2005). *Serving Youth with Microfinance: Perspectives of Microfinance Institutions and Youth Serving Organizations,* USAID.

o   Ministry of Youth and Sports. (2010*). National Youth Policy of Ghana: Towards an empowered youth, impacting positively on National Development.* Accra: national gazette.

o   Olaniyi, O.A., Adewale, J.G. (2012). *Information on Maize Production among Rural Youth: A Solution for Sustainable Food Security in Nigeria.* Library Philosophy and Practice (e-journal). Paper 724.

o   Valerie, L. (2009). *Youth in Agriculture; Challenges and Opportunities: Proceedings of the 30th Regular Meeting of the Conference of Heads of Government of the Caribbean Community, 2-5 July 2009*, Georgetown, Guyana.

o   White, B. (2012). *Agriculture and the generation problem: rural youth, employment and the future of farming. FAC . ISSER Conference\Young People, Farming and Food.* Accra, 19-21 March 2012.

o   World Bank. (2012). *Agricultural Innovation Systems: An Investment Sourcebook.* Washington, DC: World Bank.

YOUR KNOWLEDGE HAS VALUE

- We will publish your bachelor's and
  master's thesis, essays and papers

- Your own eBook and book -
  sold worldwide in all relevant shops

- Earn money with each sale

Upload your text at www.GRIN.com
and publish for free